Anderson Miguel Lenz
Welder Siena
Yuri Feruzzi

Development of an LVDT and datalogguer for phytomonitoring

Anderson Miguel Lenz
Welder Siena
Yuri Feruzzi

Development of an LVDT and datalogguer for phytomonitoring

Development of an LVDT transducer and datalogguer for phytomonitoring

ScienciaScripts

Cover image: www.ingimage.com

This book is a translation from the original published under ISBN 978-3-330-76193-3.

Publisher:
Sciencia Scripts
is a trademark of
Dodo Books Indian Ocean Ltd. and OmniScriptum S.R.L publishing group

120 High Road, East Finchley, London, N2 9ED, United Kingdom
Str. Armeneasca 28/1, office 1, Chisinau MD-2012, Republic of Moldova, Europe
Managing Directors: Ieva Konstantinova, Victoria Ursu
info@omniscriptum.com

Printed at: see last page
ISBN: 978-620-8-37938-4

SUMMARY

LENZ, Anderson Miguel and SIENA, Welder. Design and development of an LVDT *(Linear Variable Differential Transformer)* sensor and systems for: supplying, adapting and acquiring the device's signal (Industrial Maintenance Technology) - Federal Technological University of Paraná. Medianeira, 2012.

Sensing plays an important role in agriculture due to the growing demand for food and the difficulty of measuring or diagnosing behaviours that designate a problem or optimise crop development. Precision agriculture provides an evolution in this concept, as this model of cultivation involves the insertion of systems capable of estimating control parameters. However, the purpose of this work is to present the construction of an LVDT transducer, along with its operating principle and information on the development process, test modelling and the response of positioning signals from the LVDT core.

Keywords: Precision agriculture, Crop development, Electronic systems, *Phytomonitoring,*

SUMMARY

CHAPTER 1	3
CHAPTER 2	4
CHAPTER 3	5
CHAPTER 4	13
CHAPTER 5	17
CHAPTER 6	23
CHAPTER 7	29
CHAPTER 8	34
CHAPTER 9	39
CHAPTER 10	40

CHAPTER 1

INTRODUCTION

Obtaining data through sensors allows for the expansion of ways of acquiring evaluative information, which makes it possible to optimise resources and processes. Based on this trend and the need to automate processes, the aim of this work is to demonstrate the construction and analysis of a linear variable differential transformer (LVDT*)* to assess crop development, characterised as *phytomonitoring.*

According to JORAS, J.S.D. (2003), *phtyomontitoring* is one of the ways of managing irrigation properly and efficiently, where information on the soil's water conditions is required. These conditions can be obtained using an LVDT sensor that captures the radial contraction of the plant according to the conditions to which it is subjected. According to HINCKLEY T.M and BRIKERHOFF D.N. (1975), these contractions stem from the behaviour of the plant stem, which has its diameter reduced during the day due to transpiration not being compensated by water absorption, and during the night there is an increase in the stem due to water absorption and little transpiration. This variation resulting from the change in matric potential makes it necessary to use sensors to capture this movement.

According to DONGWON, Y. (2011) the use of variable reactance sensors comes from the advantages of good linearity and voltage gain, which makes it possible to measure events. The magnetic displacement device allows positions to be measured precisely, making it one of the most widespread positioning sensors used in general, due to its acquisition method, good linearity and robustness. These attributes provide the sensor with a variety of applications, such as in the agricultural and mechanical industries and in research centres SPEZIA (2011).

CHAPTER 2

OBJECTIVES

2.1 GENERAL OBJECTIVE

To develop a functional system for monitoring plant stem diameter, capable of evaluating crop development by storing data on stem diameter variation.

2.2 SPECIFIC OBJECTIVE

Studying the applicability of the LVDT sensor in agriculture and developing it in conjunction with a power supply, adaptation and signal acquisition circuit. Partitioning the LVDT system to develop and evaluate each stage, integrating them into a final project that enables the functionality of the sensor and, consequently, the analysis of crops through methodologies such as phyto-monitoring.

2.3 WORK STRUCTURE

The work was structured according to the sequence of assembly of the systems, where the theoretical background was first used as a basis for choosing the materials, the shapes and the manufacturing process of the mechanical part, after which we moved on to the development of the sensor's power supply and adjustment systems, followed by the acquisition system, and with all the previous stages completed, the field test of the sensor.

CHAPTER 3

LINEAR VARIABLE DIFFERENTIAL TRANSFORMER (LVDT)

3.1 INITIAL CONSIDERATIONS

Efficient crop management, according to YUNSEOP (2011), is a concern in many farming systems, but as a way of supporting agricultural efficiency, the use of sensors is spreading to enable control of the irrigation and development system, in order to maximise production while optimising the use of agricultural resources.

As a way of optimising resources, obtaining data from the LVDT system through stem variation has been considered a potential tool in crop management, as it provides an immediate, consistent and reliable response to the plant's water status. Control of the system generally occurs because crops, to a greater or lesser extent, respond to variable soil moisture conditions by retracting or dilating their stems. This anatomical variation caused by water deficit is analysed by contact sensors, which measure the radial variation of the stem in response to the soil's matric potential, JORAS, J.S.D. (2003).

According to CORREIA (2005), electronics is currently playing an increasingly important role in various technological areas. This is also true in agriculture, where there is an increase in the inclusion of electronic systems. This is due to the fact that these systems have many advantages over mechanical systems, such as a smaller physical size of the system, a longer lifespan and easier control.

One reason for the growing use of electronics in agriculture is the search for optimisation of resources and productivity, so sensory systems are applied to improve the agricultural process.

Sensory systems are made up of transducers which, according to MARTINO (2010), convert physical quantities into electrical signals which are then sent to the control unit corresponding to the system being measured. There are currently various types of sensors in the agricultural process for taking measurements such as position, speed, humidity and temperature. In the case of the LVDT sensor presented here, it is characterised by a displacement or position transducer.

This type of sensor has inductive characteristics based on mutual changes in inductance, self-inductance or magnetic resistance. As explained by SPEZIA (2010), this sensor is usually encapsulated in stainless steel or polymers, making it insensitive to mechanical or electrical interference. Figure 1 shows how the sensor works.

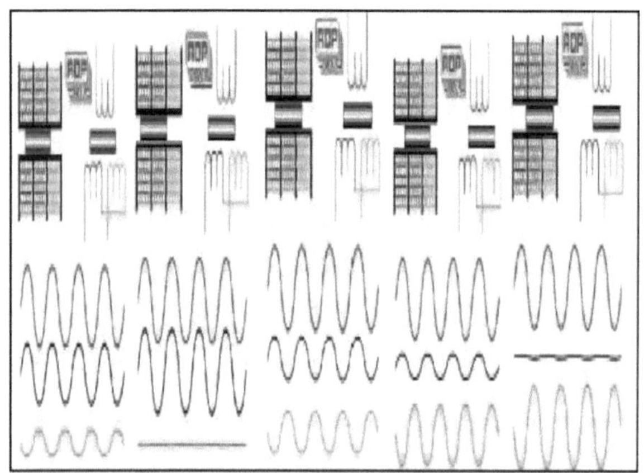

Figure 1 Showing the functional form of sense Source: Pedro Filipe Alhais Lopes

As shown in Figure 1, the red waveform and the red coil represent the alternating power supply signal, the blue waveform and the blue coil represent the 1ª secondary coil and the green coil represents the *2-coil,* so when the core represented by the grey colour is mostly inserted in the field of the 1ª coil there will be a higher signal amplitude in this coil, as shown in the second column of waveforms, when the core is positioned evenly between the two coils the two signals will be of the same value as shown in the 3ª column, and when most of the core is positioned in the *2-coil* there will be a greater amplitude of the signal in this coil, so the LVDT sensor works with output variation proportional to the positioning of the core, which gives this sensor many fields of use due to its good linearity and easy implementation MAZI (2010).

As far as its physical constitution is concerned, according to LOSITO (2011), the LVDT contains three concentric and coaxial coils surrounded by an encapsulation, the centre coil being the emitting magnetic circuit, which is powered by high-frequency alternating current. The secondary windings are located around the emitter coil. The moving part, i.e. the core, is made of a high magnetic permeability material, nickel and chrome steel, which is inserted into the coil. When the core moves from the centre position, where it results in zero output voltage, the intensity of the voltage in the secondaries in a given direction of the core's movement increases, while the opposite movement of the core causes this electrical voltage to decrease. Graph 1 shown by MEYDAN (1992) exemplifies the transducer's working characteristics according to the position of the core.

Graph 1. Characteristic response model of an LVDT sensor Source: MEYDAN (1992)

As can be seen in Graph 1, the sensor has a so-called nominal range that goes from -100% to +100% of the core position, which according to MEYDAN (1992), refers to the area where the voltage output signal is most linear. This means that when the sensor is designed, it must have a slightly larger measurement than that required. As an intrinsic characteristic of this sensor, there is a loss of linearity at the ends of the windings.

In addition to the windings, the LVDT is divided into several other elements which will be detailed below, along with the design features. These elements that make up this type of device consist of: encapsulation, spool, core, clamping, power supply, matching and signal acquisition.

3.2 ENCAPSULATION

According to CRESCINI (1995), the encapsulation of the sensor consists of protecting the magnetic circuit against mechanical shocks or inclement weather. In order to make the device lighter and more resistant, it was developed in technyl nylon material, which provides relevant physical characteristics, as the polymer used has a low density and high resistance to weathering. These requirements are fundamental when choosing the material due to the sensor's application, which requires a system that interferes as little as possible with crop development. Figure 2 shows the physical characteristics of the encapsulation developed for the sensor.

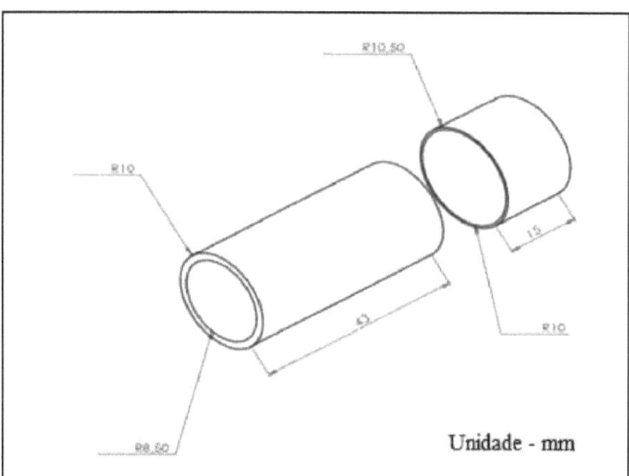

Figure 2 - Encapsulation and resin insert for connection protection.

The encapsulation seen in Figure 2 was developed on a mechanical lathe until the visualised shape was obtained.

3.3 CARRETEL

In the LVDT, the spool is the element responsible for allocating the coils. As a characteristic, according to BALBINOT (2006), this partitioning of the sensor consists of a mechanical element with a cylindrical shape and three coaxial and concentric recesses and generally of the same dimensions.

To develop the winding allocation device, the same polymer mentioned above was used to facilitate construction and reduce the sensor's mass so that it can be applied to crops in the early stages of development. Figure 3 shows the physical shape and dimensions of the reel design.

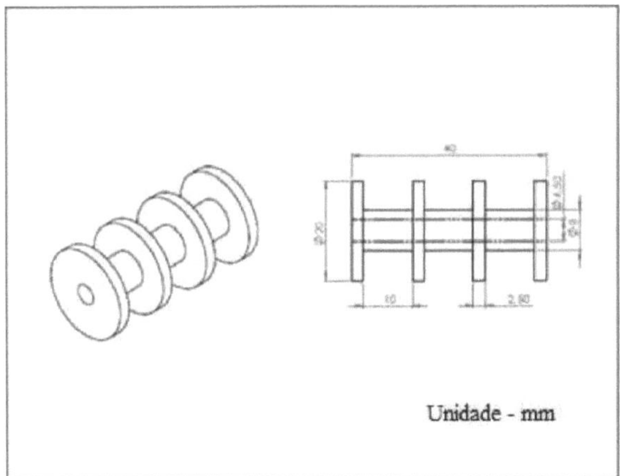

Figure 3 - Illustration of the reel element

3.4 MAGNETIC CIRCUITS - TRANSMITTER AND RECEIVER

In its physical constitution, an LVDT sensor has two magnetic circuits juxtaposed to each other. On the centrally-placed spool is the primary winding, which is the device's emitter circuit. The secondary circuit is located on the other parts of the spool, SPEZIA (2010).

As for the primary circuit, according to MARTINO (2011) its power supply can vary from 1kHz to 20kHz in alternating voltage.

work shows a gain in the magnetic density of the sensor, making it more sensitive and with greater response amplitudes.

The LVDT's secondary circuit is located around the transmitter coil, connected in series with each other, but with a winding offset of 180^{s} and spaced symmetrically in relation to the primary coil and between them. Figure

4 shows the LVTD transmitter and receiver magnetic circuit.

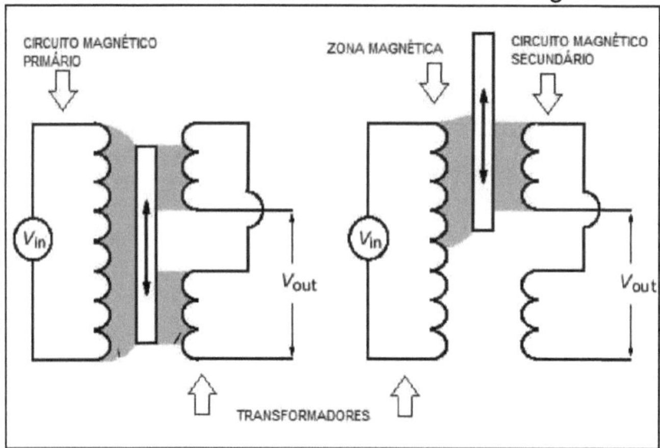

Figure 4. display of magnetic circuits Source: Meydan (1992)

The secondary part of the windings seen in Figure 4 consists of the mutual inductance element, which undergoes this phenomenon according to one of the concepts of magnetism in which the exposure of a coil in a region of pulsating magnetic flux induces a voltage in it. In the case of the sensor, the primary magnetic circuit provides the pulsating flux, due to its high-frequency AC supply. As a result, the secondary circuit is immersed in this pulsating magnetic flux and suffers the phenomenon of mutual inductance, resulting in the output signals.

3.5 COIL CONSTRUCTION CHARACTERISTICS

According to BALBINOT (2006), an LVDT sensor contains coils according to the project, knowing that sensors with higher capacities are found above 1000 coils per coil.

However, due to the application of the device developed, it cannot have large dimensions, so as seen in Figure 3, the dimensions of the reel, on which the primary and secondary windings were developed with AWG32 wire with 600 turns in each of the partitions of the reel. Figure 5 shows the final result of the reel with the magnetic circuits.

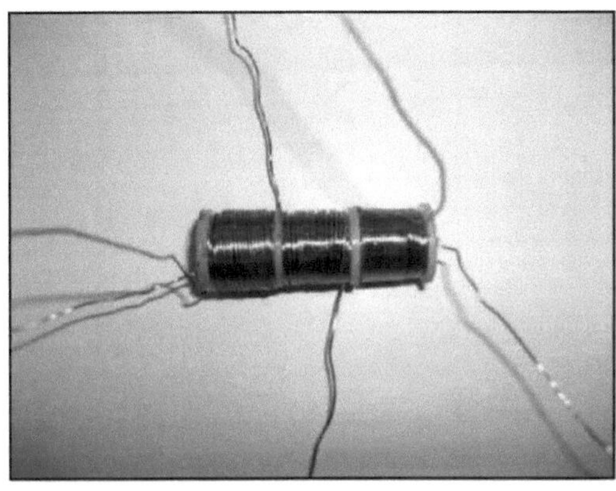

Figure 5 - LVDT reel after winding the magnetic circuits

3.6 NUCLEUS

This element is responsible for the variation in the output signal, as it causes the magnetic flux to be concentrated according to its movement, which intensifies or reduces the signal in proportion to its position; its constitution for this work was a Nickel-Chromium Steel alloy. The magnetic flux concentration can be seen in Figure 4.

3.7 FIXATION

This work presents an LVDT sensor for *phytomonitoring*, and based on this focus the sensor mount was developed in order to fulfil the coupling to plants in the most appropriate way possible. According to KANO (1998), the fixture was developed, which consists of two parts, an upper flat plate and a lower angled plate, these plates are used to fix the LVDT and couple the assembly to the plant. Figure 6 shows the characteristics of these elements.

Figure 6 - Fastening model used

As can be seen in Figure 6, the LVDT's angle plate has been designed so that it can be attached to the species (plant) with good fixation, as the element developed is a triangle. This means that the sensor has a larger contact area when compared to two flat plates. Figure 7 shows the fixing situations mentioned here.

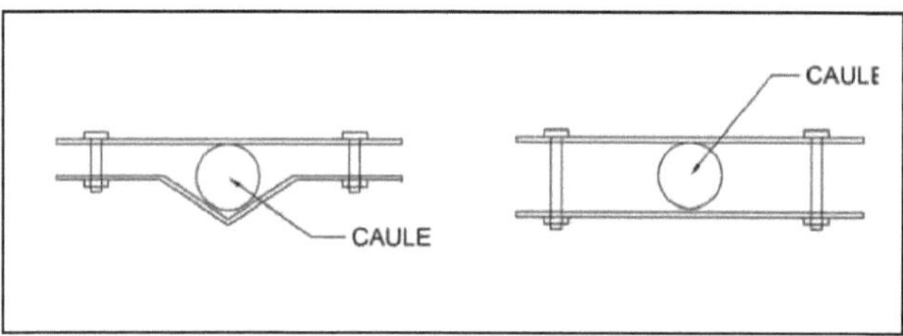

Figure 7. Comparison of fixation systems

The clamping device with an angled base appears to be the best application solution, as it is one of the most widely used types of couplings in the agricultural sensory niche.

CHAPTER 4

POWER SUPPLY CIRCUIT - EXCITATION

4.1 INITIAL CONSIDERATIONS

The power supply circuit for an LVDT sensor is based on high-frequency sinusoidal excitation, which defines this power supply as an oscillator circuit. According to SEDRA (2004), an oscillator is an amplifier stage that generates a certain frequency, conditioned by the value of its components, and keeps it within certain limits. This type of circuit uses positive feedback to compensate for the loss of power during the generation of each cycle. According to VARGAS (2002), the electrical properties of transistors allow them to be used in oscillators. For a transistor to operate in a high-frequency sinusoidal circuit, it must operate at its quiescent point, a characteristic range of amplifier operation.

There is a wide range of oscillators, such as: wien bridge oscillator, hartley oscillator, colpitts oscillator, lock oscillator, double T oscillator and the phase shift oscillator (used in the development of the sensor). The circuit is detailed below.

4.2 LVDT EXCITATION CIRCUIT - OSCILLATOR

The sensor in question was built using a transistorised phase rotation oscillator. According to VARGAS (2002), this type of oscillator operates under the reactance-capacitance relationship, where the resistances assigned to the circuit along with the capacitors cause the oscillation and phase change, producing signals that are found at frequencies of 15Hz to 20kHz. Figure 8 shows the oscillator circuit used to power the sensor.

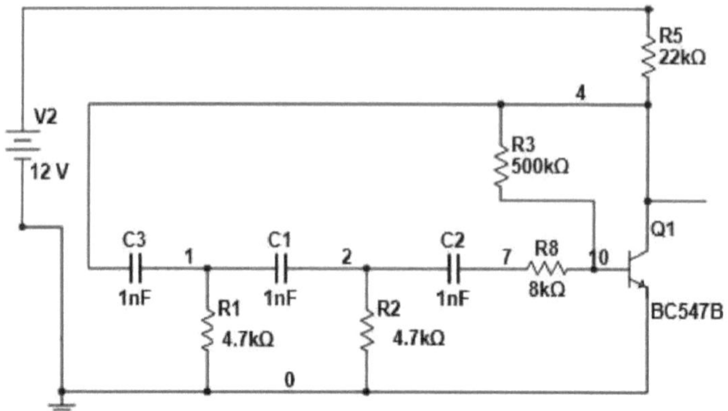

Figure 8. Oscillator circuit used to power the sensor.

The components used to make the circuit are shown in Figure 8. According to VARGAS (2002), the circuit elements can be defined according to the desired frequency using Equation 1.

$$f = \frac{1}{2\pi RC\sqrt{6}} \qquad (1)$$

f = required frequency.

R = resistance value.

C = capacitance value.

However, the oscillator circuit was built using the phase-shift oscillator circuit developed by BRAGA (1998). Despite using the circuit developed by Braga, some changes had to be made to make it compatible with the LVDT. The changes were made to the circuit's complement, where the impedance coupler and current amplifier circuits were introduced into the oscillator, in order to adapt the circuit's operation to the transducer's power supply.

4.3 LVDT EXCITATION CIRCUIT - VOLTAGE FOLLOWER

An operational amplifier, or op amp, is a very high gain differential amplifier with high input impedance and low output impedance. It is normally used in oscillator circuits, filters and instrumentation systems.

According to WENDLING (2004) one of the ways of using amplifiers is the follower configuration, which has the characteristic of unit gain, the same amplitude and polarity. The unitary follower configuration normally operates as a stage isolator, current booster and impedance coupler. The emitter configuration provides amplifiers with essential applications in numerous projects so that there is maximum energy transfer between two circuits, be they amplifiers, oscillators, etc.

Figure 9 shows two circuits (a) and (b). In circuit (a), the Vo voltage according to COUGHLIN (2001), when connected directly to the load, the internal resistance produces a resistive divider, directly affecting the voltage delivered. However, the source and load when interspersed by a follower circuit maintain the voltage.

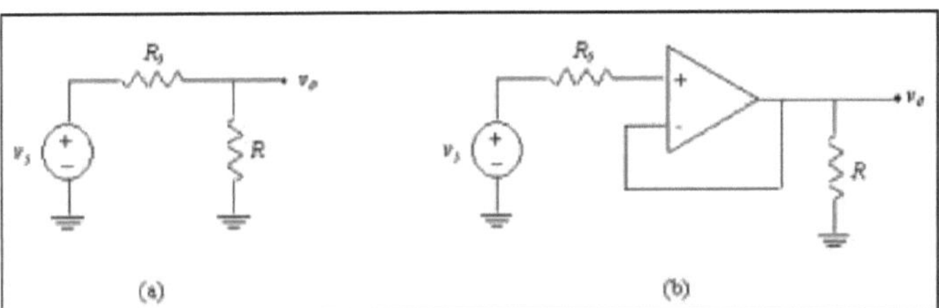

Figure 9. power transfer comparison

In order to maintain maximum power transfer between circuits, an LM324 integrated circuit was used in emitter follower configuration. Figure 10 shows the complete sensor power supply circuit.

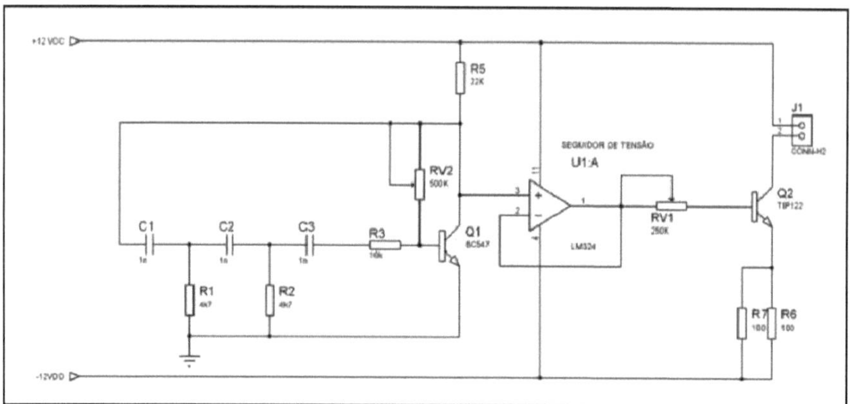

Figure 10. Excitation circuit for the transducer

As can be seen in Figure 10, the excitation circuit is made up of an oscillator circuit, a follower circuit and a current amplifier (TIP-122), inserted to provide the sensor with a higher density magnetic field. On this basis, the LVDT is inserted into the terminals of connector J1.

CHAPTER 5

SIGNAL MATCHING CIRCUIT

5.1 INITIAL CONSIDERATIONS

According to NATALE (2007), generally the electrical quantity output by a transducer is not directly manipulable. For example, the output voltage range is not desired, the signal power is low or the type of quantity is not required. For these reasons, transducers need a system that manipulates the signal in the most appropriate way, and this manipulation of the signal is called conditioning.

According to NATALE (2007), a signal conditioner is a piece of equipment that converts one electrical quantity into another, also electrical, but adapted to the specific application. In most cases, the output electrical quantity is voltage. Figure 11 illustrates the block diagram of a sensor with a conditioning system.

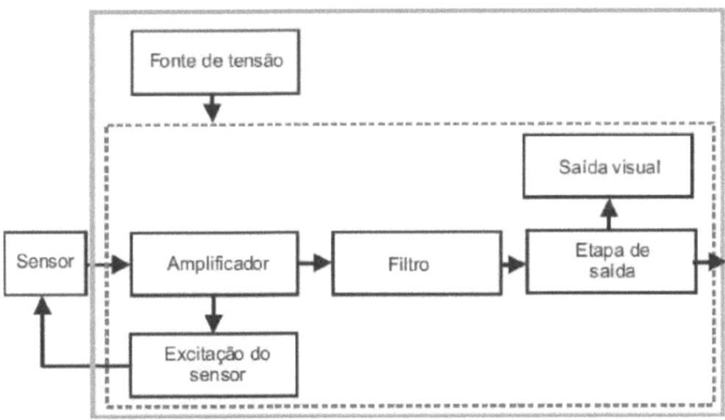

Figure 11. Block diagram for signal conditioning
Source: NATALE (2007)

As illustrated in Figure 11, the excitation system will feed the sensor and when it is influenced by the medium it is monitoring the device will emit a signal proportional to the influence suffered. The signal emitted by the sensor will pass through the amplifier, filter and finally into the acquisition system, thus generating control or evaluation measurements. The path of the signal from excitation to acquisition is called the signal conditioning system.

5.2 DEVELOPMENT CHARACTERISTICS OF THE SIGNAL MATCHING SYSTEM

For the transducer shown here, it was necessary to condition the voltage value of the secondaries so that it would be possible, using a microcontroller's analogue to digital converter, to analyse this value representing the measured stroke. The ADC *(Analogue to Digital Converter) of* the chosen microcontroller, the PIC 18F4550, is limited in range to between 0 and 5V, and in resolution to around 4.8 mV.

The output signal of the secondary windings, as a characteristic of the sensor, has the same frequency as the excitation signal, f = 9.7 kHz. This value comes from various tests, where by varying the frequency of the power supply we found that at this frequency, the output signal had the lowest noise level and amplitude suitable for the adjustment system. In order to apply a linear variation to the sensor, the electronic system for digital analogue conversion was used, thus adapting the sensor to the characteristics of the acquisition system according to the sensor's variation measurements, which represent 0 to 10 mm of displacement.

To analyse the signal conditioning system, it was subdivided into 4 blocks: coupling and amplification, rectification, filtering and subtraction, and finally offset.

5.3 COUPLING AND AMPLIFICATION SYSTEM

As already mentioned, this fragment of the system used an LM324 for the follower configuration, which is responsible for coupling the impedances of the systems, in this case the LVDT output and the signal conditioner input, as shown in Figure 12.

As a complement to this system, one of the LM324's ports was used for non-inverting amplification. This stage was used due to the low amplitude of the output provided by the LVDT, which is a millivolt signal. So the use of the amplifier raised the output to 5 volts. Figure 12 illustrates this fragment of the signal conditioning circuit.

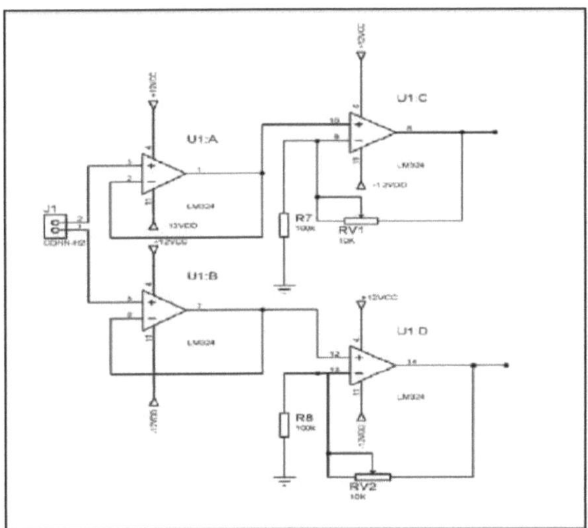

Figure 12. Coupling and amplification circuit of the signal conditioning system

As seen in Figure 12, the LVDT signal comes from connector J1. From this point on, the amplifiers in emitter follower configuration operate on each of the sensor's polarities, coupling the impedances between sensor and circuit. After coupling, the signal is amplified and then conditioned by the circuit.

5.4 RECTIFICATION, FILTERING AND SUBTRACTION SYSTEM

The system referred to in this sub-item corresponds to the rectification required for signal processing. In this process, the diode transforms the still high-frequency AC signal into a pulsating wave. Due to the characteristics of the LVDT sensor, after rectification, according to ALHAIS (2008), the graphical expression showing this action (rectification) will contain two domains. Figure

13 shows the sensor's graphical expressions according to rectification.

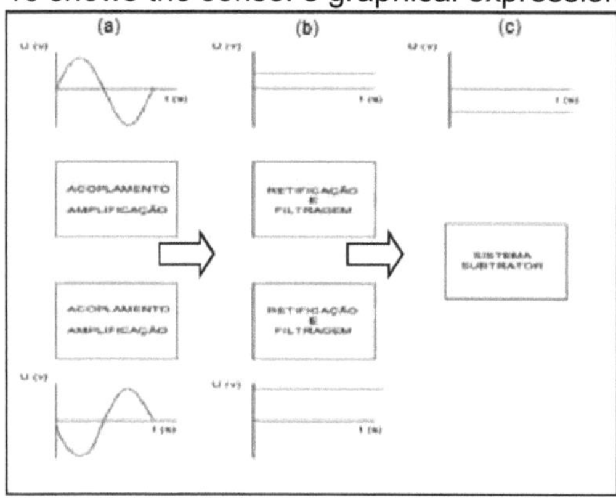

Figure 13. Illustration of the step-by-step forms of the signal processing system (conditioner) Source: Vilhais

Figure 13 shows the behaviour of the LVDT signal under the conditioner. The outputs of the sensor's secondary are visualised in the first two graphs, which show a sinuous signal with an offset of 180^2. As already discussed, this signal is amplified and then rectified, so the subtraction circuit is used. This circuit consists of a typical assembly of a non-inverting subtractor. According to ALHAIS (2008), the subtractor amplifier consists of a configuration that makes it possible to obtain a voltage signal at the output equal to the difference between the applied signals. Figure 14 shows the subtractor configuration of an operational amplifier.

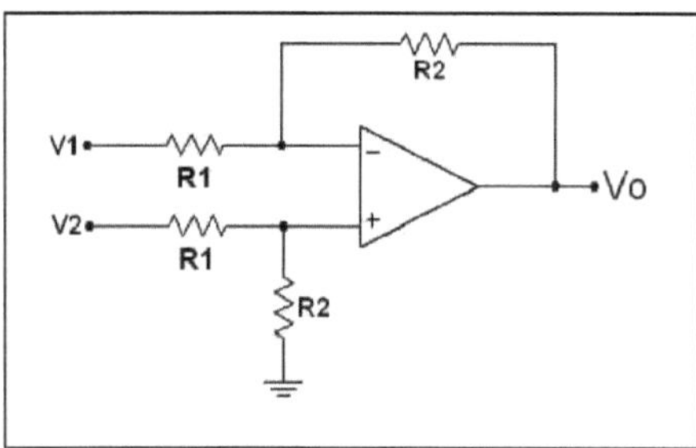

Figure 14. Sub-tractor configuration Source: Wedling

The output of a circuit in subtractor configuration has the characteristic as shown in Equation 2 by WEDLING (2010), where the voltage V_o depends on the ratio between Ri and R_2 integrated to a product of the signal difference added to the positive and negative ends of the operational amplifier.

$$V_O = \frac{R_2}{R_1}(V_2 - V_1) \qquad [2]$$

From this configuration, the graph (c) in Figure 13 is obtained, in which the voltages that formed two images are subtracted into a single signal that varies only in the positive cycle based on the offset that will be detailed below.

5.5 OFFSET SYSTEM, AMPLIFICATION AND FILTERING.

This partitioning of the signal conditioning system, the offset, is necessary, according to ALHAIS (2008), because the subtraction results in a negative signal for one half of the displacement. As the acquisition system will be carried out by a microcontroller, it does not allow negative voltage signals to be read, resulting in the need to offset the difference value.

As far as amplification is concerned, as the difference in amplitude is very small, the signal must be amplified to read the ADC and to provide a wider voltage range across the entire stroke. In the case of the sensor, this amplification results in a maximum voltage value of 3.3 volts, dimensioned to avoid possible damage to the microcontroller port due to voltage spikes. Figures 15 and 16 show the overall signal conditioning circuit, which includes the oscillation, coupling, amplification, rectification, filtering, subtraction and offset circuits.

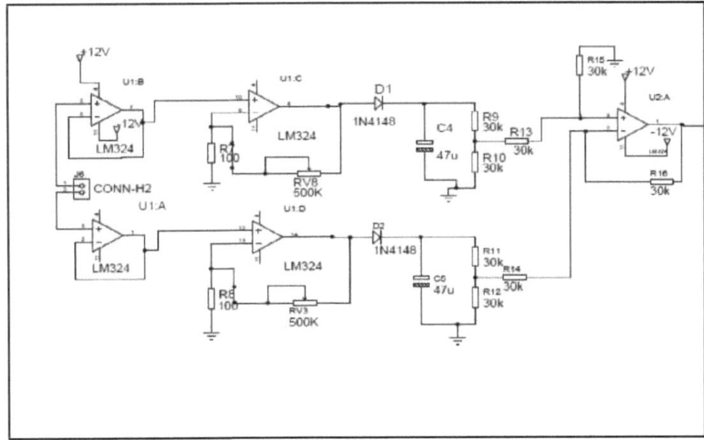

Figure 15. Signal matching circuit

Figure 15 shows the configuration starting with the signal sent by the LVDT, which is processed using an impedance coupler, amplifier, rectifier; voltage divider (to adjust the voltage) and subtractor. After these steps, the signal is directed to the rest of the circuit responsible for adapting the signal.

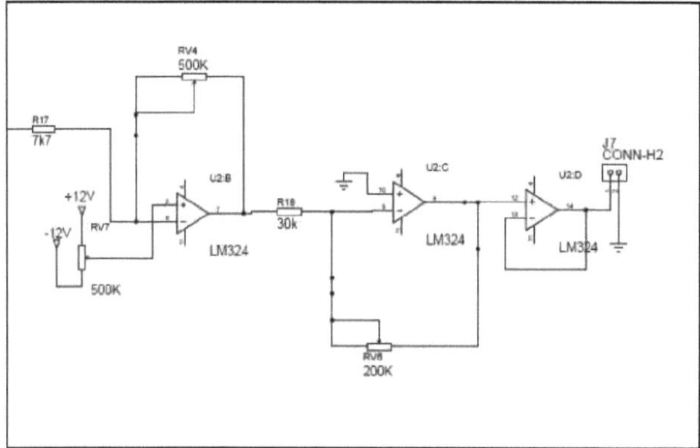

Figure 16. Signal matching circuit complement

Figures 15 and 16 represent the signal adaptation circuit in which, after all the signal processing in figure 15, it enters the circuit in figure 16 to complete the processing required to use the signal in the *datalogger*.

CHAPTER 6

DATA ACQUISITION SYSTEM

6.1 INITIAL CONSIDERATIONS

The growing evolution in the fields of process management and control has led to the development of IT for agricultural purposes over the decades, making it possible to use precision tools for analysing and acquiring data. According to PARK (2003), in all types of technology and science, data acquisition is an essential activity. The purpose of the acquisition system is to present the observer with the values of the variables being measured. Thus, the performance of a given crop in terms of fertigation and irrigation is widely studied in its various configurations, with the aim, among other things, of achieving a balanced performance and reducing operating costs. Based on these premises, this chapter will detail the sensor acquisition system, a *datalogger* which consists of a device that receives data from one or more sensors, processes this data and stores it in a digital format. The system in question will be developed on the PIC 18F4550 microcontroller.

6.2 5.2 PIC 18F4550 MICROCONTROLLER

Microcontrollers are characterised internally by a computer system, which includes a CPU *(Central Processor Unit)*, data and program memory, a clock system, I/O (Input/Output) ports, as well as other possible peripherals. According to MIYADAIRA (2011), the 18F4550 microcontroller consists of a system based on the Harvard architecture where the processor accesses data memory and instructions via two different buses. The instruction set of this type of machine (PIC18F4550) is based on RISC architecture (computer with reduced instruction set). This machine is also characterised by being 8-bit with 32 kbytes of program memory and 2,048 kbytes of RAM (Randon access memory). The power supply of this device can vary between 4 and 5.5 Volts and operates at frequencies of up to 48 MHz, or 12 MIPS (Million Instructions Per Second).

The model used has 40 pins, 35 of which can be configured as 1/O's (input - output). Various peripherals can be used in this microcontroller, such as: 256-byte EEPROM or other capacity, two CCP (Capture, Compare or PWM) and ECCP (Enhanced, Capture, Compare or PWM) modules, thirteen analogue and digital (A/D) modules with 10-bit resolution and programmable acquisition time, two analogue comparators, EUSART communication, an 8-bit TIMER and more.

6.3 DEVELOPMENT OF THE SAD - DATA ACQUISITION SYSTEM

The acquisition system was developed using circuit configurations according to MIYADAIRA (2011), which can be seen in Figure 17.

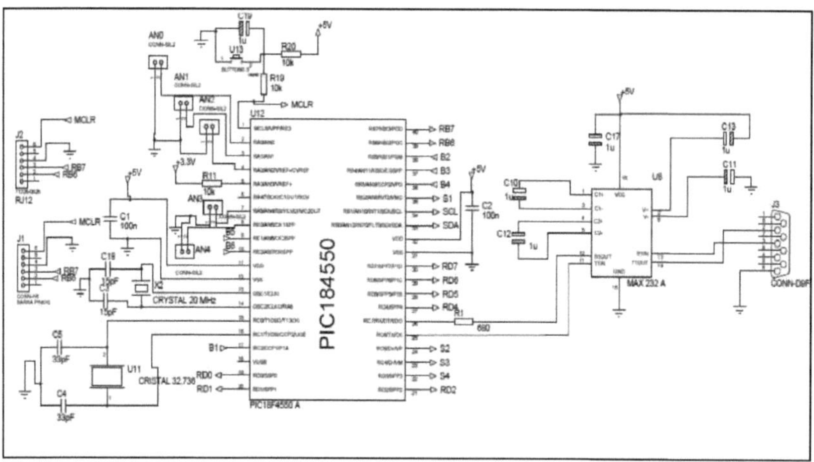

Figure 17. *Hardware* configurations for developing the datalogger
Source: Miyadaira (2011)

In the configuration shown in figure 17, you can see the peripherals needed to use the PIC 18F4550, such as the oscillator crystal, noise-inhibiting capacitors coupled to the microcontroller's Vcc and Vss ports, the Max 232 integrated circuit for serial communication and the terminals used to access the analogue channels (AN1, AN2, AN3, AN4).

As can be seen in the schematic of the circuit used, the *datalogger*'s *hardaware* contains 4 CMOS Serial EEPROM memories of 256 kbytes, with I2C™ communication, as shown in figure 18.

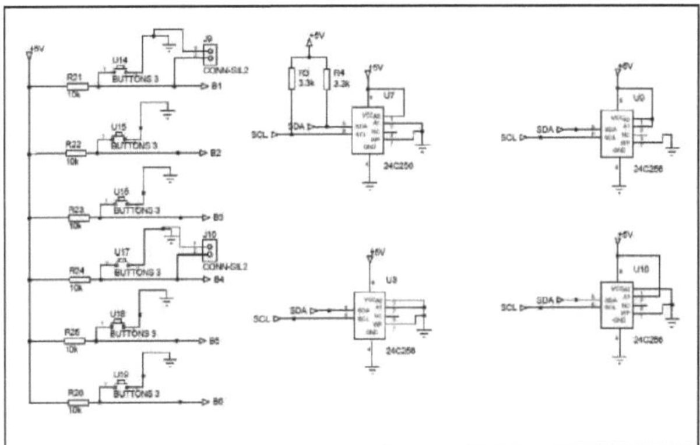

Figure 18. *Hardware* configuration for data memories

Other *hardware* features of the *datalogger* are: 4 opto-coupled relay outputs Figure 19; 2x16 lcd display; 5 analogue inputs; and access to CCP1 and CCP2.

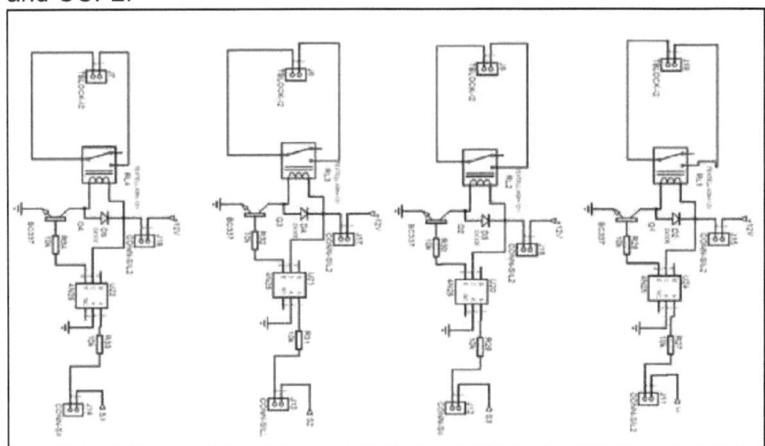

Figure 19. Opto-coupled outputs

If the core is exposed to the primary's magnetic field for a long time, it will lose its high magnetic permeability characteristic, as its internal domains will align permanently, thus becoming a magnet. To avoid this, relays were added to the *hardware* to trigger the LVDT 15 minutes before acquisition, which is enough time to magnetise the core and take the measurement, so being exposed to the magnetic flux for half the time will increase its useful life, and it will consume half the energy compared to operating continuously. Figure 19 shows the schematic of the four relay outputs for driving LVDT 15. The aim of four relays comes from future applications where the same *datalogger* will acquire data from four LVDTs. *The in-circuit* mode (1CSP - *In-Circuit Serial Programming)* was chosen to record the programme on the microcontroller, which is more practical as there is no need to remove the microcontroller from the electronic circuit to record the chip. All the features presented so far are demonstrated in a final result of the *datalogger hardware* project, which is visualised in figure 20.

Figure 20. Datalogger *hardware*

6.4 SOFTWARE DEVELOPMENT

According to COELHO (2009), *software* development requires stages that precede this process, which can be organised in different ways, such as texts and comments, using tools such as dictionaries, diagrams and flowcharts. A flowchart shows geometric figures with a brief description of the process, lines and arrows describing the sequence of activities, showing the path of information in a structured way, helping to maintain organisation in process documentation and programming. In developing the *datalogger*, a flowchart of the programme's structure was used, which is shown in Appendix A.

After defining the monitoring behaviour (flowchart), the *software* was developed. The *hardware* manipulation system *(software)* was developed in ANSI standard C language. The type of language was defined according to JÚNIOR (1999), who states that the main advantage of using high-level languages (in this case the C language) is that it allows the designer to interact less with the *hardware when it* comes to controlling it (setting register banks and initialisation sequences). In this way, the designer dedicates his time basically to the logic of the problem and not to the internal details of the chip.

To develop the C programme, Microchip's MPLAB programming environment was used. This platform provides access to the C18 compiler, which transforms high-level language into machine language, manages the project, *debugs* and writes to the chip. Figure 21 shows the development environment used.

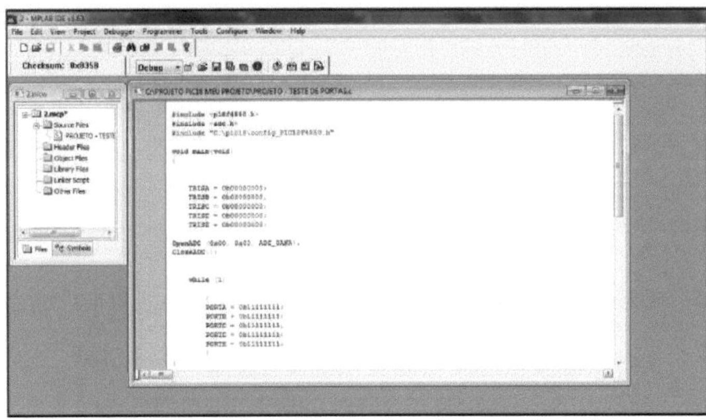

Figure 21. Microchip MPLAB development environment

Figure 21 shows the *software* work area where the control project for recording on a microcontroller was created.

In addition to the programme creation *software*, the *datalogger* used various modules available in PIC *hardware*. To develop the control, we basically used the Analogue-Digital Converter, I^2C communication for writing to EEPROM memory, TIMERs, 2X16 LCD display, and the use of all these modules was based on MIYADAIRA (2011).

The main function of the program was to record the voltage values read by the AD1 analogue channel in an external EEPROM, with times defined by the configuration of a global variable called "timehere".

From the *software* menu, you can edit various programme parameters, and check the programme's operating status indicators, such as whether it is on or off, the voltage value currently being read, and the reading of the data stored in the EEPROM. For more information on the operating logic, see Appendix A for the *Dendrometer* 1.0 *datalogger* software.

CHAPTER 7

SENSOR CALIBRATION TEST

7.1 INITIAL CONSIDERATIONS

A good quality measurement system (SM) must be able to operate with small errors. Its construction and operating principles must be designed to minimise systematic and random errors throughout its measurement range, under its nominal operating conditions. However, no matter how good the characteristics of an MC are, it will always present errors, either due to internal factors or to the action of externally influenced quantities. Perfect characterisation of the uncertainties associated with these errors is of great importance if the measurement result is to be reliably estimated. Although in some cases the errors of a measurement system can be analytically or numerically estimated, in practice experimental procedures are used almost exclusively. Through the experimental procedure known as calibration, it is possible to correlate the values indicated by the measuring system and their correspondence with the quantity being measured. This chapter therefore presents, in general terms, the characteristic aspects of LVDT sensor calibration.

7.2 CALIBRATION METHODOLOGY

Once the assembly, oscillator and conditioning have been built on a printed circuit board, the sensor connected to the final circuit is characterised. The characterisation is only done at this stage so that the signals can be read directly by the acquisition system, thus obtaining the actual values measured by the LVDT, with all the impedances involved coupled.

In the test in question, the displacement per output voltage relationship was extracted by measuring the voltage per step. This characterisation of the sensor's calibration was carried out in accordance with BALBINOT(2006), in which one way of acquiring and surveying the calibration line is to submit the sensor to a static support, so that its core can be moved by a high-precision measuring device, such as a micrometer. The items needed to calibrate the sensor are shown in Table 1.

Table 1. Elements used to calibrate the transducer

Items	Function
sheet metal structure	sensor bracket
micrometre	core displacement
feeding system	sensor power supply
signal acquisition system	variation reading

The calibration procedure was carried out using the elements shown in Table 1, where the variation in the voltage signal was checked by moving the micrometer every 0.5mm. The signals coming from the sensor's output were then read by the microcontroller's ADC channel, which showed the voltage values on the display using the programming logic.

Figure 22 shows the physical assembly of the methodology described by BALBINOT (2006) for obtaining the calibration line of the LVDT sensor. Based on this methodology, the device was subjected to a coupling of 10V peak-to-peak (system power supply) and a frequency of 10kHz (Kilo-Hertz), all values supplied by the power supply system. The values were read off the developed *datalogger*. The figure in question shows the LVDT between the metal structure and the micrometer responsible for the calibration variation (red arrow) above it (structure indicated by the blue arrow).

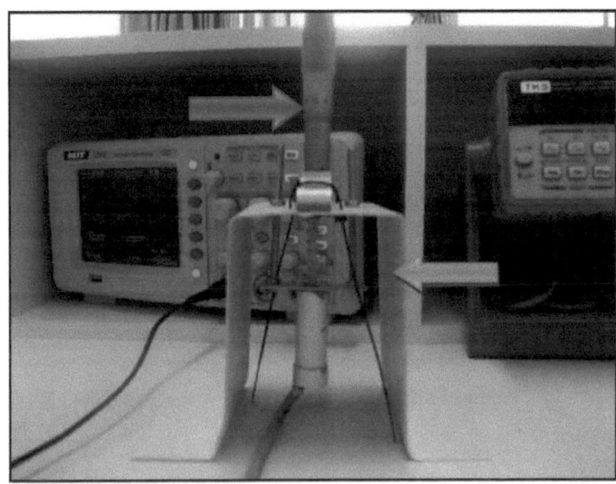

Figure 22. Structure designed to measure the sensor

Using the methodology applied, tests were carried out on the sensor to draw the calibration line. Graph 2 shows one of the repetitions with data beyond the linear region from 1 to 9 mm, showing the bulging or loss of linearity of the output voltage at the ends, which shows that the sensor has LVDT characteristics. Repetitions 1, 2, 3 and 4 refer to the analyses of voltage as a function of displacement. These are the results of the four tests carried out on the sensor, and graph 3 shows the 4 tests superimposed to evaluate the behaviour of hysteresis, which would result in significant measurement errors depending on its magnitude.

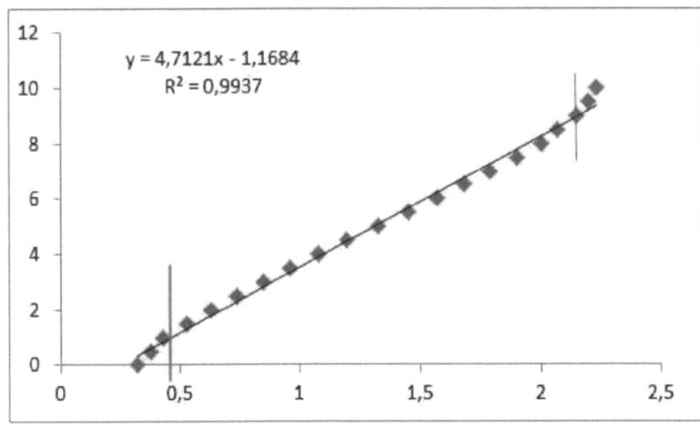

Graph 2. Repetition with extreme non-linear regions

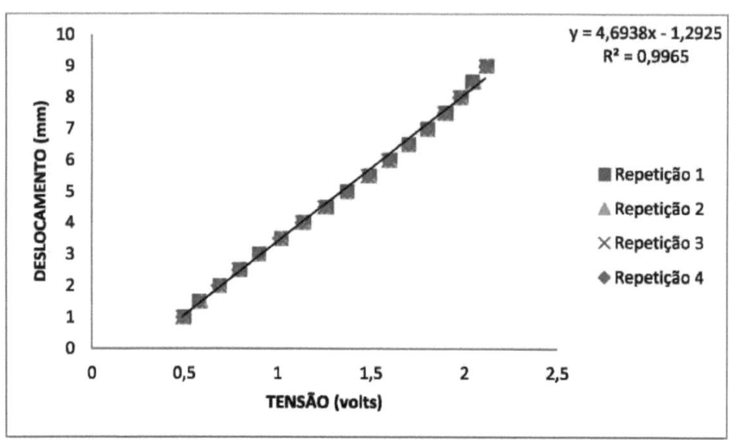

Graph 3. plotting analyses

Graph 3 was used to check for hysteresis in the sensor due to the residual magnetic charge in the core. By looking at the points, it is possible to see that there is almost no difference between one point and another, and there is no hysteresis.

After carrying out the four repetitions, the sensor's straight line equation was obtained, which returns the displacement that occurred as a result of the voltage variation. Equation 3 obtained a Pearson correlation coefficient of 0.996, which indicates a highly reliable correlation between tension and displacement. Equation 3 is shown below:

$$Y(mm) = 4{,}6938x(V) - 1{,}2925 \qquad [3]$$

Once the equation had been obtained, the average percentage error between the value predicted by the equation applied to the voltage read by the microcontroller and the known displacement was calculated. The calculation resulted in an average error of 1.69%. Table 2 shows the values calculated by Equation 3.

Table 2. Percentage error verification data

Error checking

Tension	displacement	y Forecast (mm)	Error	|Error (mm)	Error %
0,49	1,0	1,007	0,007	0,007	0,7
0,6	1,5	1,524	0,024	0,024	1,6
0,68	2,0	1,899	-0,101	0,101	5,0
0,79	2,5	2,416	-0,084	0,084	3,4
0,92	3,0	3,026	0,026	0,026	0,9
1,01	3,5	3,448	-0,052	0,052	1,5
1,13	4,0	4,011	0,011	0,011	0,3
1,24	4,5	4,528	0,028	0,028	0,6
1,36	5,0	5,091	0,091	0,091	1,8
1,45	5,5	5,514	0,014	0,014	0,2
1,53	6,0	5,889	-0,111	0,111	1,8
1,67	6,5	6,546	0,046	0,046	0,7
1,8	7,0	7,156	0,156	0,156	2,2
1,89	7,5	7,579	0,079	0,079	1,1
1,97	8,0	7,954	-0,046	0,046	0,6
2,05	8,5	8,330	-0,170	0,170	2,0
2,11	9,0	8,611	-0,389	0,389	4,3
			Average error		1,69 %

As can be seen in Table 2, the errors ranged from 0.2 to 5% error, and the average error was 1.69%. This shows that in the linear region of the sensor, which ranges from 1 to 9mm, the sensor will give reliable measurements, which is adequate for a good measurement.

CHAPTER 8

EVALUATIONS OF THE LVDT MONITORING SYSTEM APPLIED TO A SUNFLOWER.

In order to test the equipment in the field, the decision was made to install it on a fast-growing, commercial type of plant whose stem was not of the stalk type, which is easily damaged. Based on these conditions, we decided to use sunflower.

The crop chosen is highly adaptable to various climates, tolerates low temperatures and is relatively drought-resistant.

These characteristics make sunflower highly adaptable for off-season cultivation in the Centre-West region, or as a successor to off-season maize in the South, according to CAMPOS (2007).

The cultivation of sunflower provides derivatives for its use, such as animal feed, oil extraction and its use in the production of biofuels, according to a study carried out by (SAVANACHI, 2008).

Sunflower seeds were planted in a container at a depth of 5 cm on 1 February 2013, and after a certain period of vegetation, the LVDT was installed in the stem. Before installing the transducer, the initial diameter of the stem was measured, which came out at 12.26 mm.

The sensor was installed with a support at the back, as shown by the red arrow in Figure 23, to prevent the plant from tipping over, and also to adjust the angle of contact of the sensor stem. This ensures that the sensor is always in contact with the crop forming a 90° angle, avoiding unwanted inclinations which would result in plant deformation and error in measuring the variation in the stem.

Figure 23. Support structure for evaluating young crops

As far as field installation is concerned, power was supplied from a computer supply, which when connected to the mains provides the +12V and -12V needed to run the system. As the sensor operates in the field and is subject to varying weather conditions, the device was enclosed in polymeric protection to safeguard it. Figure 24 shows how the *datalogger* is accommodated *in the* field.

Figure 24. System used to secure the sensor for field operation

After installing the sensor in the field, data collection began on 28/02/2013 at 9am. The configuration established in the *software* was a monitoring analysis carried out every 30 min, enabling daily monitoring of the sunflower's behaviour. Consequently, on the sixth day of collection, the following graph of stem variation was obtained. This shows the crop's variation in relation to the soil's matric state, a key index for making irrigation decisions.

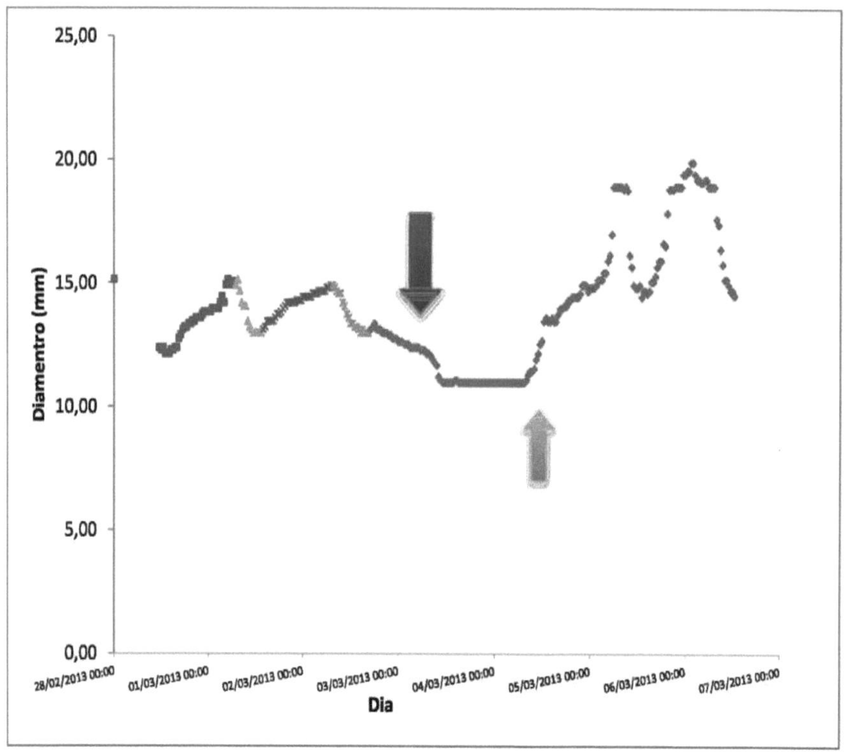

Graph 4. Variation in stem diameter according to monitoring

Graph 4 shows that on the first two days, the plant's stem varied normally throughout the day, contracting during the day due to exposure to the sun and expanding at night. In the region indicated by the red arrow, we can see a reduction in the variation in daily amplitude, which is characterised by water *stress*, where the plant no longer has enough water to maintain its metabolism. This characteristic is confirmed in Figure 25, where the soil is at a reduced matric potential, which causes a direct water deficit for the plant.

Figure 25. Plant in a state of water stress

After visually observing and monitoring the deficit in crop development, the sunflower soil was wetted with water on 04/03/2013.

The restructuring of the sunflower can be seen in the green arrow in Graph 6, where the crop once again showed a greater daily variation in the stem, showing its recovery from the deficit in the soil's matric potential.

The maintenance of the soil led to an increase in humidity, which resulted in the sunflower recovering. This recovery of the plant's condition can be seen in Figure 26.

Figure 26. Sunflower restructuring.

CHAPTER 9

CONCLUSION

After building and testing the sensor, it was found to be a potential tool to be used to check the water status of crops and their development. As has been shown, the sensor developed had adequate ranges in terms of error rate, its mass resulted in a total of 50 grams and the coupling mode in the tests carried out showed no damage to the crop under study, thus enabling the transducer to be used as a monitoring tool.

Despite the sensor's full operation, one of the device's constructive features proved to be extremely necessary during development. The feature in question refers to the sensor's central tap, which initially didn't exist, making it impossible to adjust the signal due to the lack of a tap and consequently the lack of a reference in the adjustment circuit, making it impossible to develop. In the new project, the central tap was integrated, making it possible to adjust the signal in the transducer and therefore coupling it to the *datalogger.*

As for the sensitivity of the transducer, it proved to be suitable for detecting micrometric variations in stem diameter. Thus, according to the variation and subsequent data acquisition, the sensor quantitatively shows the state of development, which it characterises in the monitoring, and on this basis decisions can be indexed to irrigate if the monitored crop has water senility, or decisions can be made to cease irrigation in order to acquire a particular characteristic provided by the crop in the state of water stress. An example of a crop to which a state of stress can be attributed is the grape, which, when it is in a state of stress, releases a maximised sucrose content in its metabolism. This characteristic is required for the production of wines and jams.

CHAPTER 10

REFERENCES

ALHAIS, P.L. **Sensor System for a Formula Student Competition Car.** Dissertation for a Master's Degree in Electrical and Computer Engineering, Universidade Técnica de Lisboa, 2008

BALBINOT, **A. Instrumentation and measurement fundamentals.** Rio de Janeiro LTC Vol.2 2007.

BRAGA, N. **Banco de circuitos** 3ed. São Paulo, 1998.

CAMPOS, M. L. **Indications for the use of sunflower**, technical communication, EMBRAPA 2007

CARR, Joseph J. Elements of electronic instrumentation and measurement. Pearson

1996. P 340-343(BELT)

Coelho HS. **Software documentation: a necessity.** Texto Livre: Linguagem e Tecnologia. 2009. (2):1 1-6.

Coughlin, F. Driscoll,F. **Operational Amplifiers and Linear Integrated Circuits** 4e Ed. Person(2001)

Crescini, D., A. Flammini, D. Marioli, and A. Taroni, **"Application of an FFT based algorithm to signal Processing of LVDT position sensors,"** *IEEE Trans. Instrum. Meas,* vol. 47, no. 5, pages 1119-1123, Oct. 1998.

DONGWON, Y. SANGYONG, H. JUNGHO, P. SONAM, Y. **Analysis and design of lvdt.** Sth International Conference on Ubiquitous Robots and Ambient Intelligence (U RAI 20), Korea, November 2011.

HINCKLEY T.M and BRIKERHOFF D.N. **The effects of drought on water relations and stem shrinkage of quercus alba,** 1975.

KANO, Y. HASEBE, S. HAUNG, C. **New Type Lvdt Position Detector.** IEEE, International Conference on Ubiquitous Robots and Ambient Intelligence, 1998.

LOSITO, R. ET ALL. **Design of a Linear Variable Differential Transformer With High Rejection.** IEEE transactions on magnetics, vol. 46, no. 2, february 2010.

MASI and R. LOSITO, **LHC collimators low levei control system,** IEEE

Trans. Nucl. Sei., vol. 55, no. 1, pages 333-340, Feb. 2008.

MARTINO, M. ET ALL. **Study of magnetic interference on a lvdt prototype.** IEEE transactions on magnetics, vol. 46, no. 2, february 2010.

MEYDAN, T. and HEALEY G.W. **Linear Variable Differential Transformer - linear displacement transducer utilising ferromagnetic amorphus metallic glas ribbons.** IEEE sensors and actuators, vol 32,1992.

MIYADAIRA, A.N. **Microcontroladores PIC18, aprenda [e]programame em linguagem C.** 2[e]ed. São Paulo-SP Ed. Érica, 2011.

NATALE. F. **Automação Industrial,** série brasileira de tecnologia, 7[a]ed., São Paulo Editora Érica.(2007)

Park, J., Mackay, S., **Practical Data Acquisition for Instrumentation and Control Systems,** Elsevier, 2003

SEDRA, A.S. **MICROELECTRONIC CIRCUITS,** by Oxford University press, Inc 2004

SPEZIA, G. **Automatic test bench for measurement of magnetic interference on lvdts.** IEEE transactions on instrumentation and measurement, vol. 60, no. 5, may 2011.

WENDLING, M. **Operational Amplifiers**, Universidade Estadual Paulista (Unesp), 2010. Available in:http://www2.feg.unesp.br/Home/PaginasPessoais/ProfMarceloWendling/3-operational-amplifiers-v2-0.pdf

YUNSEOP, K EVANS, R.G.; IVERSEN, W.M. **Remote sensing and control of an irrigation system using a distributed wireless Sensor network.** IEEE transactions on instrumentation and measurement, vol. 57, no. 7, july 2008.

VARGAS. **Techniques with digital systems.** 2002. Available at: http://www.vargasp.net/download/livros/tecnicas_digitais.pdf

JÚNIOR, V. **C language for** PIC **microcontrollers**. VIDAL Personalised Projects 1999.

SAVANACHI, E. **Os girassóis da Petrobrás**, dinheiro rural Edição 50 Agronegócios, 2008

JORAS, J.S.D. **Evaluation of the use of sap flow and variation in stem and branch diameter in the determination of citrus water conditions, based on irrigation management.** Dissertation to obtain the degree of doctor of agronomy. USP 2003.

APPENDIX A - *DATALOGGER* FLOW CHART

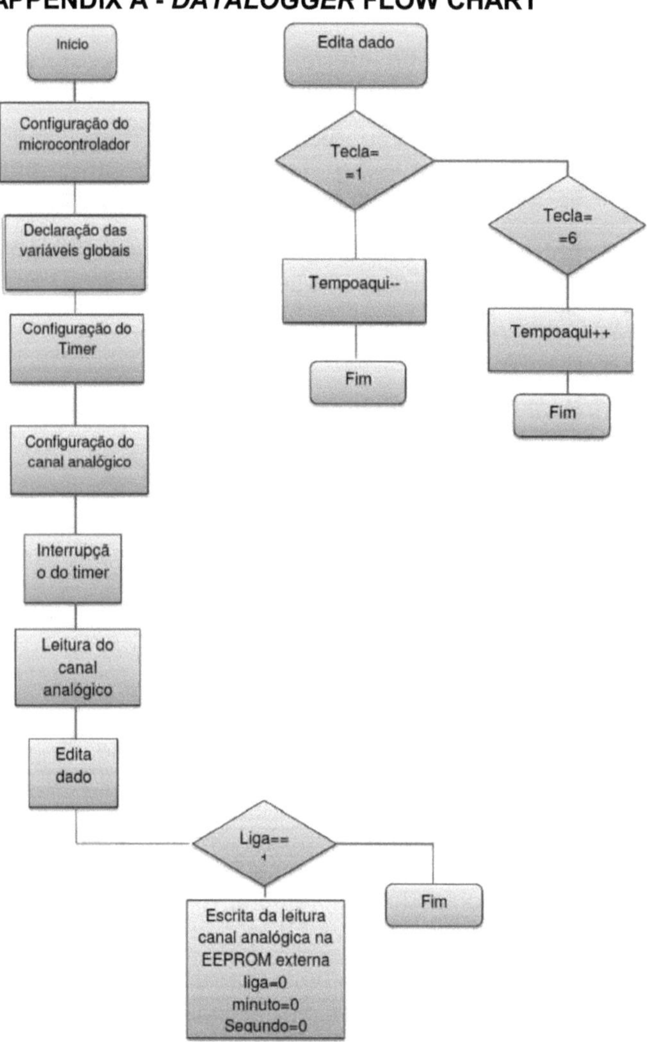

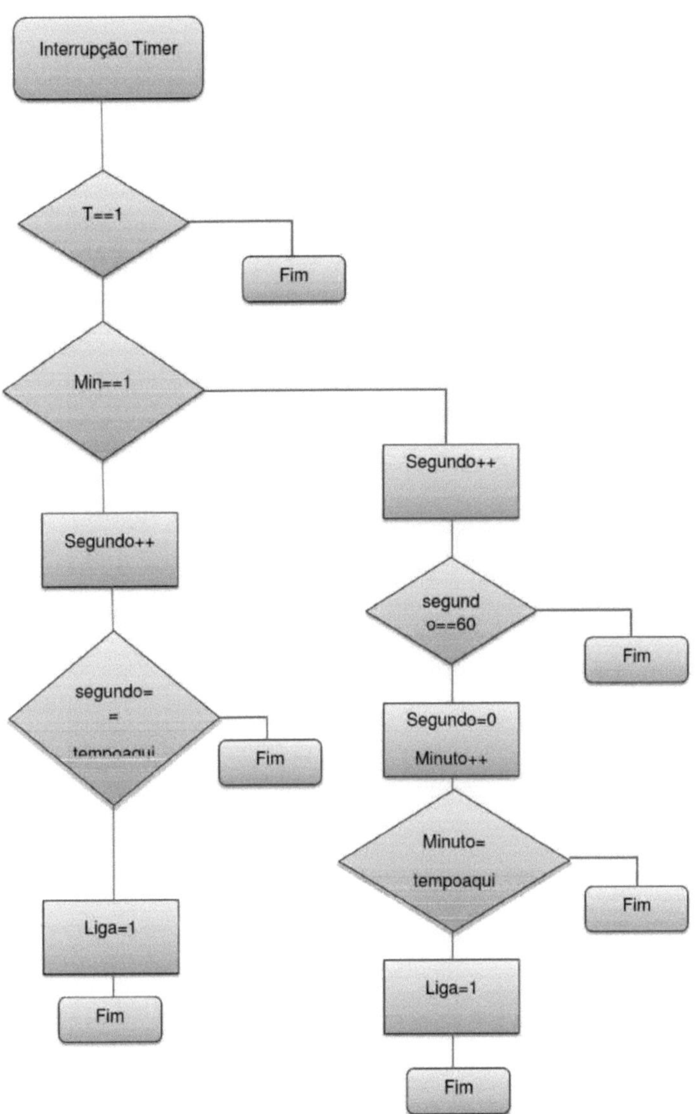
Interrupção Timer
T==1
Fim
Min==1
Segundo++
Segundo++
segund
o==60
Fim
segundo=
=
Fim
Segundo=0
Minuto++
Minuto=
tempoaqui
Fim
Liga=1
Fim
Liga=1
Fim

Printed by Books on Demand GmbH, Norderstedt / Germany